WHAT IS PALAEONTOLOGY?

Palaeontology is the study of ancient plants and animals that lived millions of years ago. They are studied by looking at **fossils** that have been preserved underground.

The scientists who study fossils are called **PALAEONTOLOGISTS.**

IS THIS THE BIGGEST DINOSAUR EVER?

DISCOVER THE SCIENCE BEHIND **PALAEONTOLOGY**
(pay-lee-on-TOH-luh-jee)

Written by Olivia Watson

Illustrated by Verónika Cháves Morales

Words that are tricky to understand are in **bold**. Find out what they mean in the glossary.

Words that are difficult to say are in *italics*. Find out how to say them at the back of the book.

Way down at the bottom of South America is a place called Patagonia. It's a huge land with tall mountains, sandy **deserts**, and wide farms. One day, a shepherd was hunting for a lost sheep when something incredible happened...

What was this mysterious bone? The shepherd got help from *palaeontologists*, scientists who spend their days working with fossils. When they uncovered the fossil bone, they were excited to see that it was from a group of huge dinosaurs called **titanosaurs** – the biggest dinosaurs to walk the Earth!

Titanosaurs are also from a group of plant-eating dinosaurs called **sauropods**. They had incredibly long necks and tails, and walked – very slowly – on four legs. These **prehistoric** giants were so huge that they spent most of their time eating just to stay alive!

The shepherd's discovery was only the start! The scientists found many more fossils in the same place. They began to feel that this was something very special.

The palaeontologists were amazed when two giant leg bones were uncovered. They believed they might have found a very big dinosaur indeed!

The scientists very carefully and slowly dug up more than **200 FRAGILE FOSSILS!**

When the bones were taken to the lab, they were able to work out amazing things, like how this huge dinosaur moved, how heavy it was, and even when it lived.

Tests on the fossils showed that this enormous titanosaur lived **101.6 MILLION YEARS AGO!**

This period of time is called the **Cretaceous** period.

So the titanosaur roamed Earth at the same time as the *Velociraptor*, *Triceratops*, and *T. rex*, but they didn't all live in the same place!

The **dig site** got bigger and bigger, and the scientists made another important discovery – pieces of
SAUROPOD EGGS!

One **unique** egg had the patterns of skin inside. It was the first time in history that a human could see what dinosaur skin looked like!

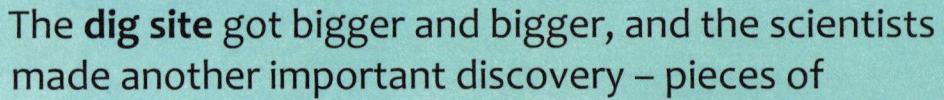

The scientists knew that fossils are too **fragile** to build a full-size dinosaur **skeleton**, so they scanned the bones one by one…

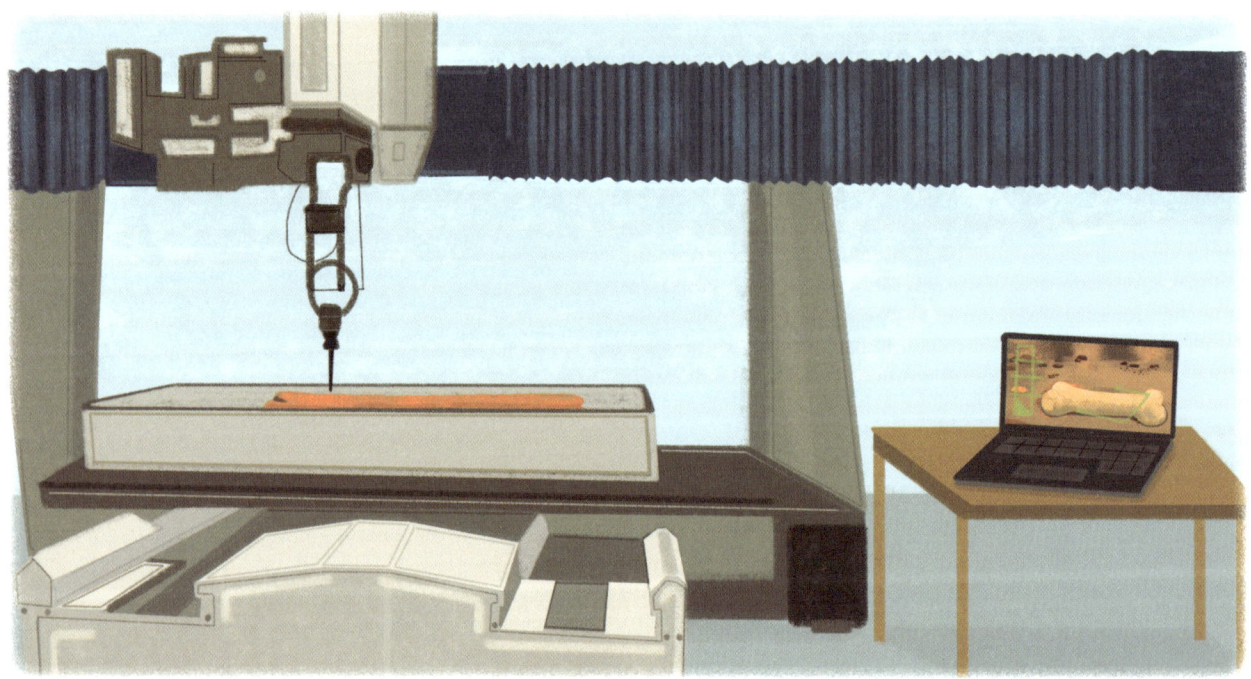

and used these scans to make a 3D computer model of the skeleton. They sent this to a special factory that makes copies of bones…

then, with the world's biggest puzzle pieces, they fitted together the replica bones to build the whole skeleton!

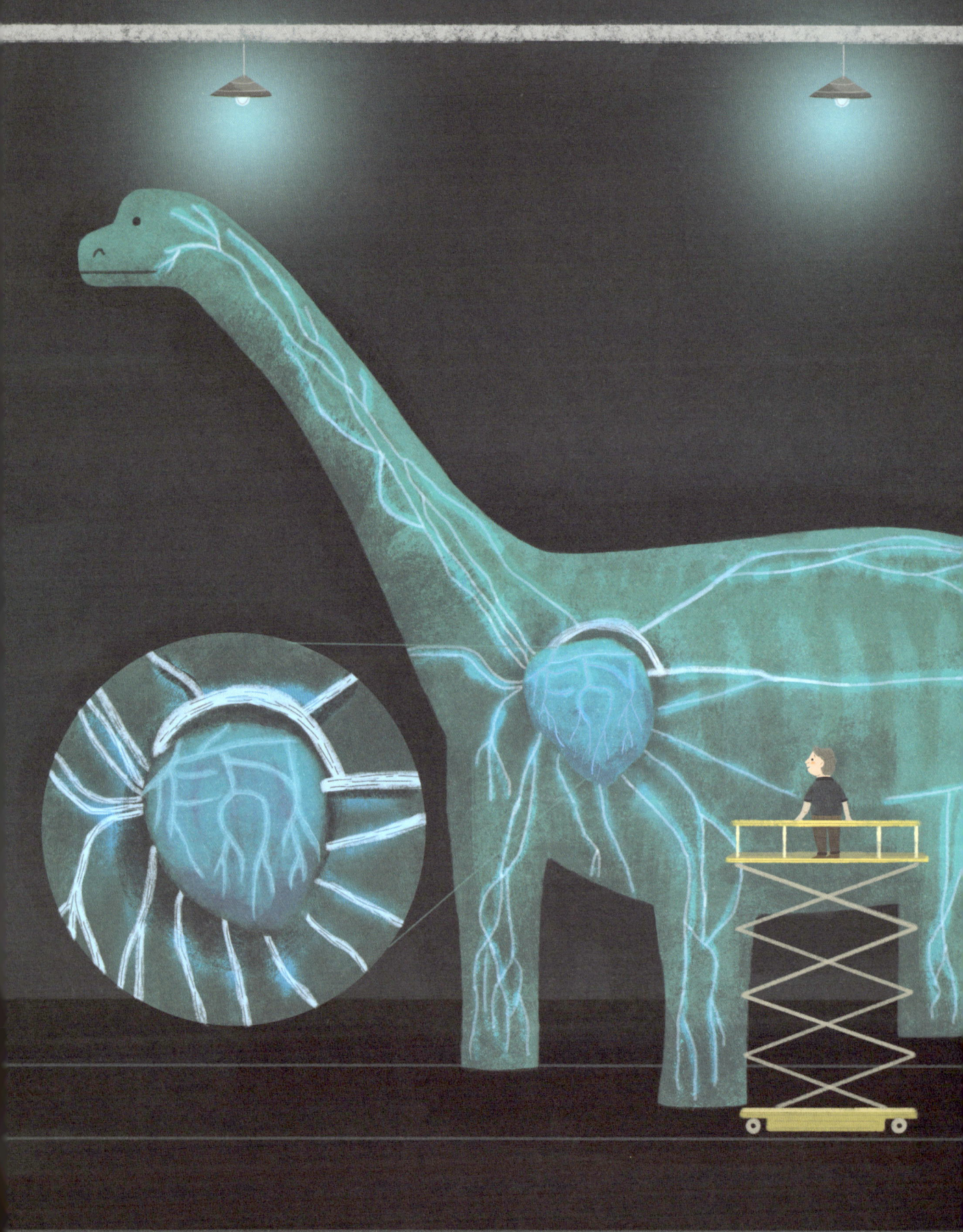

The scientists used the 3D computer model to understand how the dinosaur lived. This titanosaur's body was so huge that it needed a gigantic heart, weighing as much as a motorcycle. The heart pumped blood around its body and massive tail muscles to help it move its heavy legs.

They worked out from the bones that the dinosaur was about 37.5 metres (123 feet) long – the same as three buses! So this huge **herbivore** needed to eat a lot to survive. The trees kept growing taller to avoid being eaten, so the sauropods grew taller too!

Titanosaurs could reach food from the tallest treetops just like giraffes can today.

The scientists also thought about the dinosaur's teeth. Plant-eaters' teeth were small and flat, but meat-eaters' teeth were huge and sharp! A tooth from *Tyrannotitan*, a **predator** even bigger than T. rex, was found near a titanosaur bone with a bite mark in it! Perhaps titanosaurs were dinner for this meat-eater?

After years of work, the palaeontologists showed their discovery to the world! Since they had uncovered so many fossilised secrets, they knew this was a **NEW TYPE OF DINOSAUR!**

They called it the *Patagotitan Mayorum*.

The palaeontologists could now imagine what this titanosaur looked like when it was alive and compare its size with other animals. They finally knew for sure – they had found

THE BIGGEST DINOSAUR EVER!

Tyrannosaurus rex

Record-breaking
DINOSAURS

The Patagotitan mayorum may have won the title of biggest dinosaur ever found, but palaeontologists have discovered lots of other dinosaurs with record-breaking features too...

The smallest dinosaur was... COMPSOGNATHUS!

For years, chicken-sized *Compsognathus* was considered the smallest dinosaur. However, recent discoveries suggest that even smaller ones may have existed!

The longest neck belonged to...
MAMENCHISAURUS!

Mamenchisaurus' neck is estimated to have been 15 metres (49 feet) long.

The biggest tooth belonged to...
TYRANNOSAURUS REX!

The longest T.rex tooth ever found measured a terrifying 30 centimetres (12 inches) from root to tip!

The longest tail belonged to... DIPLODOCUS!

Diplodocus and Supersaurus had the longest tails of all. They are thought to have been over 14 metres (45 feet) long!

Dinosaur fossils are being discovered all the time, so palaeontologists are always updating records.

Amazing
FOSSIL FACTS

There's so much to discover about the world of palaeontology. Do you know the answers to some of the world's biggest questions about dinosaurs?

WHEN WAS THE FIRST DINOSAUR FOUND?

The first record is from the 1670s. Dr Robert Plot wrote about a Megalosaurus thighbone that he thought belonged to a giant human!

Megalosaurus

DO PALAEONTOLOGISTS ONLY FIND DINOSAURS?

NO! Some palaeontologists study prehistoric plant life or **microfossils** instead.

WERE ALL DINOSAURS SCALY?

NO! In the 1990s, many fossils which showed signs of feathers were uncovered. Up until then, palaeontologists had thought all dinosaurs were scaly!

WHERE DID THE WORD 'DINOSAUR' COME FROM?

'Dinosaur' means 'terrible lizard'. It was invented in the 1840s when palaeontologist Sir Richard Owen named some fossils he believed belonged to a group of huge **reptiles** that no longer existed.

WHEN WERE MOST BONES DISCOVERED?

During the Bone Wars in the 1800s, Edward Drinker Cope and Othniel Charles Marsh competed to find the most fossils. Between them, they discovered 142 dinosaur **species**!

GLOSSARY

Cretaceous – a period of time that lasted from about 145 to 66 million years ago.

Deserts – places that are extremely dry.

Dig site – an area of land where something specific is being dug up.

Fossils – the remains of plants and animals that lived long ago.

Fragile – something that breaks easily.

Herbivore – an animal that only eats plants.

Microfossils – fossils that are so small they can only be seen with a microscope.

Predator – an animal that hunts other animals for food.

Prehistoric – the time before humans existed.

Reptiles – a group of cold-blooded animals, including snakes, lizards, and dinosaurs.

Sauropods – a type of huge dinosaur that had a long neck and tail, trunk-like legs, but a small head.

Skeleton – the bony frame that supports and protects the body of a person or animal.

Species – a group of living things that share characteristics and features. For example, T.rex and Triceratops are different dinosaur species.

Titanosaurs – a type of sauropod that had a huge and heavy body. *Need help saying this? Look below!*

Unique – something that is one of its kind; unlike all others.

HOW DO I SAY?

Compsognathus
comp-SOG-nath-us

Mamenchisaurus
ma-men-key-SORE-us

Palaeontologists
pay-lee-on-TOH-luh-jists

Palaeontology
pay-lee-on-TOH-luh-jee

Patagotitan mayorum
pat-ah-go-TIE-tan
may-OR-um

Titanosaurs
tie-TAN-oh-sores

T.rex
TEE-rex

Triceratops
try-SER-ah-tops

Tyrannotitan
tie-RAN-oh-tie-tan

Velociraptor
vel-OH-si-rap-TER

THE BIG QUESTIONS ANSWERED

This is more than just a series of books; it is a complete resource.
Accompanying each book is a variety of FREE material to engage curious kids with science.

www.thebigquestionsanswered.com

Use the QR code to visit the website, download free resources, and discover other books in the series.

On the website, find out incredible things about palaeontologists, including what they do, some of their greatest discoveries, and what it takes to become an expert in this field of science.

The material is also available for home or classroom use, supporting all the information in this book.

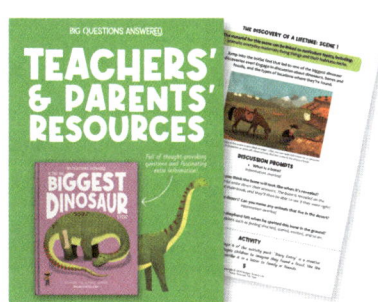

Teachers' & Parents' Resources
With discussion prompts and questions, extra information, and facts around key topics.

Young Palaeontologists' Activity Pack
Fun activities for wannabe dinosaur hunters, including creative writing, drawing, word searches, and much, much more.

BEETLE BOOKS

Beetle Books is an imprint of Hungry Tomato Ltd.

First published in 2024 by Hungry Tomato Ltd
F15, Old Bakery Studios, Blewetts Wharf, Malpas Road,
Truro, Cornwall, TR1 1QH, UK.

ISBN 9781835691243

Copyright © 2024 Hungry Tomato Ltd

No part of this publication may be reproduced, stored in a retrieval system, or transmitted in any form or by any means, electronic, mechanical, photocopying, recording, or otherwise, without prior written permission of the copyright owner.

A CIP catalog record for this book is available from the British Library.

With thanks to:
Editor: Holly Thornton
Editor: Millie Burdett
Senior Designer: Amy Harvey
Tim Cook for his valued contribution
The team at Beehive Illustration

Printed and bound in China.

Picture Credits:
(t = top, b = bottom, m = middle, l = left, r = right)
Shutterstock: Ananami 35tl; Dotted yeti 32mr; Mark_Kostrich 33ml; Michael Rosskothen 34mr; Michal Ninger 34bl; Orla 35mr; Rafael Trafanivc 35bl; Stockphoto mania 32ml.